Copyright ©

All rights reserved. No part of this publication may be reproduced, distributed, or transmitted in any form or by any means, including photocopying, recording, or other electronic or mechanical methods, without the prior written permission of the publisher, except in the case of brief quotations embodied in critical reviews and certain other noncommercial uses permitted by copyright law

Table of Contents

INTRODUCTION 3
    Applications of hydraulic systems 6
    Advantages and Disadvantages of Hydraulic system 11
    Open loop and closed loop circuits 29
    Five basic types of load-sensing systems 35
    Hydraulic pump 36
    Hydraulic fluid 43
What Is a Hydraulic System? 49
    Hydraulic Circuits 51
    Hydraulic Motors 52
    Hydraulic Cylinders 53
    What is a hydraulic system used for? 56
    Hydraulic System Components 59
Hydraulic Fluids 63
    Ten Steps to Check Optimum Viscosity Range 67
    Consolidating Hydraulic Fluids 69
    How Does a Hydraulic System Work? 71
    Open Vs Closed Hydraulic Systems 81
    CONCLUSION 98

INTRODUCTION

Hydraulic system

The controlled movement of parts or a controlled application of force is a common requirement in the industries. These operations are performed mainly by using electrical machines or diesel, petrol and steam engines as a prime mover. These prime movers can provide various movements to the objects by using some mechanical attachments like screw jack, lever, rack and pinions etc. However, these are not the only prime movers.

The enclosed fluids (liquids and gases) can also be used as prime movers to provide controlled motion and force to the objects or substances. The specially designed enclosed fluid systems can provide both linear as well as rotary motion. The high magnitude controlled force can also be applied by using these systems. This kind of enclosed fluid based systems using pressurized incompressible liquids as transmission media are called as hydraulic systems. The hydraulic system works on the principle of Pascal's law which says that the pressure in an enclosed fluid is uniform in all the directions. The Pascal's Law. The force given by fluid is given by the multiplication of pressure and area of cross section. As the pressure is same in all the direction, the smaller piston feels a smaller force and a large piston feels a large force. Therefore, a large force can be generated with smaller force input by using hydraulic systems.

The hydraulic systems consists a number of parts for its proper functioning. These include storage tank, filter, hydraulic pump, pressure regulator, control valve, hydraulic cylinder, piston and leak proof fluid flow pipelines. The schematic of a simple hydraulic system

- A movable piston connected to the output shaft in an enclosed cylinder
- Storage tank
- Filter
- Electric pump
- Pressure regulator
- Control valve
- Leak proof closed loop piping.

The output shaft transfers the motion or force however all other parts help to control the system. The storage/fluid tank is a reservoir for the liquid used as a transmission media.

The liquid used is generally high density incompressible oil. It is filtered to remove dust or any other unwanted particles and then pumped by the hydraulic pump. The capacity of pump depends on the hydraulic system design. These pumps generally deliver constant volume in each revolution of the pump shaft. Therefore, the fluid pressure can increase indefinitely at the dead end of the piston until the system fails. The pressure regulator is used to avoid such circumstances which redirect the excess fluid back to the storage tank.

The movement of piston is controlled by changing liquid flow from port A and port B.

The cylinder movement is controlled by using control valve which directs the fluid flow.

The fluid pressure line is connected to the port B to raise the piston and it is connected to port A to lower down the piston. The valve can also stop the fluid flow in any of the port.

The leak proof piping is also important due to safety, environmental hazards and economical aspects. Some accessories such as flow control system, travel limit control, electric motor starter and overload protection may also be used in the hydraulic systems

Applications of hydraulic systems

The hydraulic systems are mainly used for precise control of larger forces. The main applications of hydraulic system can be classified in five categories:

1. Industrial: Plastic processing machineries, steel making and primary metal extraction applications, automated production lines, machine tool industries, paper industries, loaders, crushes, textile machineries, R & D equipment and robotic systems etc.

2 Mobile hydraulics: Tractors, irrigation system, earthmoving equipment, material handling equipment, commercial vehicles, tunnel boring equipment, rail equipment, building and construction machineries and drilling rigs etc.

3 Automobiles: It is used in the systems like breaks, shock absorbers, steering system, wind shield, lift and cleaning etc.

4 Marine applications: It mostly covers ocean going vessels, fishing boats and navel equipment.

5 Aerospace equipment: There are equipment and systems used for rudder control, landing gear, breaks, flight control and transmission etc. which are used in airplanes, rockets and spaceships.

Hydraulic Pump

The combined pumping and driving motor unit is known as hydraulic pump. The hydraulic pump takes hydraulic fluid (mostly some oil) from the storage tank and delivers it to the rest of the hydraulic circuit. In general, the speed of pump is constant and the pump delivers an equal volume of oil in each revolution. The amount and direction of fluid flow is controlled by some external mechanisms. In some cases, the hydraulic pump itself is operated by a servo controlled motor but it makes the system complex. The hydraulic pumps are characterized by its flow rate capacity, power consumption, drive speed, pressure delivered at the outlet and efficiency of the pump. The pumps are not 100% efficient. The efficiency of a pump can be specified by two ways. One is the volumetric efficiency which is the ratio of actual volume of fluid delivered to the maximum theoretical volume possible. Second is power efficiency which is the ratio of output hydraulic power to the input mechanical/electrical power. The typical efficiency of pumps varies from 90-98%.

The hydraulic pumps can be of two types:

- Centrifugal pump

- Reciprocating pump

Centrifugal pump uses rotational kinetic energy to deliver the fluid. The rotational energy typically comes from an engine or electric motor. The fluid enters the pump impeller along or near to the rotating axis, accelerates in the propeller and flung out to the periphery by centrifugal force. In centrifugal pump the delivery is not constant and varies according to the outlet pressure. These pumps are not suitable for high pressure applications and are generally used for low-pressure and high-volume flow applications. The maximum pressure capacity is limited to 20-30 bars and the specific speed ranges from 500 to 10000. Most of the centrifugal pumps are not self priming and the pump casing needs to be filled with liquid before the pump is started.

The reciprocating pump is a positive plunger pump. It is also known as positive displacement pump or piston pump. It is often used where relatively small quantity is to be handled and the delivery pressure is quite large. The construction of these pumps is similar to the four stroke engine. The crank is driven by some external rotating motor. The piston of pump reciprocates due to crank rotation. The piston moves down in one half of crank rotation, the inlet valve opens and fluid enters into the cylinder.

In second half crank rotation the piston moves up, the outlet valve opens and the fluid moves out from the outlet. At a time, only one valve is opened and another is closed so there is no fluid leakage. Depending on the area of cylinder the pump delivers constant volume of fluid in each cycle independent to the pressure at the output port.

Pump Lift

In general, the pump is placed over the fluid storage tank. The pump creates a negative pressure at the inlet which causes fluid to be pushed up in the inlet pipe by atmospheric pressure. It results in the fluid lift in the pump suction. The maximum pump lift can be determined by atmospheric pressure and is given by pressure head as given below:

Theoretically, a pump lift of 8 m is possible but it is always lesser due to undesirable effects such as cavitation. The cavitation is the formation of vapor cavities in a liquid.

The cavities can be small liquid-free zones ("bubbles" or "voids") formed due to partial vaporization of fluid (liquid). These are usually generated when a liquid is subjected to rapid changes of pressure and the pressure is relatively low. At higher pressure, the voids implode and can generate an intense shockwave.

Therefore, the cavitation should always be avoided. The cavitation can be reduced by maintaining lower flow velocity at the inlet and therefore the inlet pipes have larger diameter than the outlet pipes in a pump. The pump lift should be as small as possible to decrease the cavitation and to increase the efficiency of the pump.

Pressure Regulation

The pressure regulation is the process of reduction of high source pressure to a lower working pressure suitable for the application. It is an attempt to maintain the outlet pressure within acceptable limits. The pressure regulation is performed by using pressure regulator. The primary function of a pressure regulator is to match the fluid flow with demand. At the same time, the regulator must maintain the outlet pressure within certain acceptable limits.

When the valve V1 is closed and V2 is opened then the load moves down and fluid returns to the tank but the pump is dead ended and it leads to a continuous increase in pressure at pump delivery. Finally, it may lead to permanent failure of the pump.

Therefore some method is needed to keep the delivery pressure P1 within the safe level. It can be achieved by placing pressure regulating valve V3. This valve is closed in normal conditions and when the pressure exceeds a certain limit, it opens and fluid from pump outlet returns to the tank via pressure regulating valve V3. As the pressure falls in a limiting range, the valve V3 closes again.

When valve V1 is closed, the whole fluid is dumped back to the tank through the pressure regulating valve. This leads to the substantial loss of power because the fluid is circulating from tank to pump and then pump to tank without performing any useful work. This may lead to increase in fluid temperature because the energy input into fluid leads to the increase in fluid temperature. This may need to the installation of heat exchanger in to the storage tank to extract the excess heat. Interestingly, the motor power consumption is more in such condition because the outlet pressure is higher than the working pressure.

Advantages and Disadvantages of Hydraulic system

Advantages

- The hydraulic system uses incompressible fluid which results in higher efficiency.

- It delivers consistent power output which is difficult in pneumatic or mechanical drive systems.

- Hydraulic systems employ high density incompressible fluid. Possibility of leakage is less in hydraulic system as compared to that in pneumatic system. The maintenance cost is less.

- These systems perform well in hot environment conditions.

Disadvantages

- The material of storage tank, piping, cylinder and piston can be corroded with the hydraulic fluid. Therefore one must be careful while selecting materials and hydraulic fluid.

- The structural weight and size of the system is more which makes it unsuitable for the smaller instruments.

- The small impurities in the hydraulic fluid can permanently damage the complete system, therefore one should be careful and suitable filter must be installed.

- The leakage of hydraulic fluid is also a critical issue and suitable prevention method and seals must be adopted.

- The hydraulic fluids, if not disposed properly, can be harmful to the environment.

Hydraulic Pumps

1. Classification of Hydraulic Pumps

These are mainly classified into two categories:

A. Non-positive displacement pumps

B. Positive displacement pumps.

A. Non-Positive Displacement Pumps

These pumps are also known as hydro-dynamic pumps. In these pumps the fluid is pressurized by the rotation of the propeller and the fluid pressure is proportional to the rotor speed. These pumps can not withstanding high pressures and generally used for low-pressure and high-volume flow applications. The fluid pressure and flow generated due to inertia effect of the fluid. The fluid motion is generated due to rotating propeller.

These pumps provide a smooth and continuous flow but the flow output decreases with increase in system resistance (load). The flow output decreases because some of the fluid slip back at higher resistance. The fluid flow is completely stopped at very large system resistance and thus the volumetric efficiency will become zero. Therefore, the flow rate not only depends on the rotational speed but also on the resistance provided by the system. The important advantages of non-positive displacement pumps are lower initial cost, less operating maintenance because of less moving parts, simplicity of operation, higher reliability and suitability with wide range of fluid etc. These pumps are primarily used for transporting fluids and find little use in the hydraulic or fluid power industries.

Centrifugal pump is the common example of non-positive displacement pumps. Details have already discussed in the previous lecture.

B. Positive displacement pump

These pumps deliver a constant volume of fluid in a cycle. The discharge quantity per revolution is fixed in these pumps and they produce fluid flow proportional to their displacement and rotor speed. These pumps are used in most of the industrial fluid power applications. The output fluid flow is constant and is independent of the system pressure (load). The important advantage associated with these pumps is that the high-pressure and low-pressure areas (means input and output region) are separated and hence the fluid cannot leak back due to higher pressure at the outlets. These features make the positive displacement pump most suited and universally accepted for hydraulic systems. The important advantages of positive displacement pumps over non-positive displacement pumps include capability to generate high pressures, high volumetric efficiency, high power to weight ratio, change in efficiency throughout the pressure range is small and wider operating range pressure and speed. The fluid flow rate of these pumps ranges from 0.1 and 15,000 gpm, the pressure head ranges between 10 and 100,000 psi and specific speed is less than 500.

It is important to note that the positive displacement pumps do not produce pressure but they only produce fluid flow. The resistance to output fluid flow generates the pressure. It means that if the discharge port (output) of a positive displacement pump is opened to the atmosphere, then fluid flow will not generate any output pressure above atmospheric pressure. But, if the discharge port is partially blocked, then the pressure will rise due to the increase in fluid flow resistance. If the discharge port of the pump is completely blocked, then an infinite resistance will be generated. This will result in the breakage of the weakest component in the circuit. Therefore, the safety valves are provided in the hydraulic circuits along with positive displacement pumps. Important positive displacement pumps are gears pumps, vane pumps and piston pumps. The details of these pumps are discussed in the following sections.

Gear Pumps

Gear pump is a robust and simple positive displacement pump. It has two meshed gears revolving about their respective axes. These gears are the only moving parts in the pump.

They are compact, relatively inexpensive and have few moving parts. The rigid design of the gears and houses allow for very high pressures and the ability to pump highly viscous fluids. They are suitable for a wide range of fluids and offer self-priming performance.

Sometimes gear pumps are designed to function as either a motor or a pump. These pump includes helical and herringbone gear sets (instead of spur gears), lobe shaped rotors similar to Roots blowers (commonly used as superchargers), and mechanical designs that allow the stacking of pumps. Based upon the design, the gear pumps are classified as:

• External gear pumps

• Lobe pumps

• Internal gear pumps

• Gerotor pumps

Generally gear pumps are used to pump:

- Petrochemicals: Pure or filled bitumen, pitch, diesel oil, crude oil, lube oil etc.

- Chemicals: Sodium silicate, acids, plastics, mixed chemicals, isocyanates etc.

- Paint and ink

- Resins and adhesives

- Pulp and paper: acid, soap, lye, black liquor, kaolin, lime, latex, sludge etc.

- Food: Chocolate, cacao butter, fillers, sugar, vegetable fats and oils, molasses, animal food etc.

External gear pump

The external gear pump consists of externally meshed two gears housed in a pump case. One of the gears is coupled with a prime mover and is called as driving gear and another is called as driven gear. The rotating gear carries the fluid from the tank to the outlet pipe. The suction side is towards the portion whereas the gear teeth come out of the mesh. When the gears rotate, volume of the chamber expands leading to pressure drop below atmospheric value. Therefore the vacuum is created and the fluid is pushed into the void due to atmospheric pressure. The fluid is trapped between housing and rotating teeth of the gears. The discharge side of pump is towards the portion where gear teeth run into the mesh and the volume decreases between meshing teeth. The pump has a positive internal seal against leakage; therefore, the fluid is forced into the outlet port. The gear pumps are often equipped with the side wear plate to avoid the leakage. The clearance between gear teeth and housing and between side plate and gear face is very important and plays an important role in preventing leakage. In general, the gap distance is less than 10 micrometers. The amount of fluid discharge is determined by the number of gear teeth, the volume of fluid between each pair of teeth and the speed of rotation. The important drawback of external gear pump is the unbalanced side load on its bearings. It is caused due to high pressure at the outlet and low pressure at the inlet which results in slower speeds and lower pressure ratings in addition to reducing the bearing life. Gear pumps are most commonly used for the hydraulic fluid power applications and are widely used in chemical installations to pump fluid with a certain viscosity.

Lobe Pump

Lobe pumps work on the similar principle of working as that of external gear pumps.

However in Lobe pumps, the lobes do not make any contact like external gear pump. Lobe contact is prevented by external timing gears located in the gearbox.

Similar to the external gear pump, the lobes rotate to create expanding volume at the inlet. Now, the fluid flows into the cavity and is trapped by the lobes. Fluid travels around the interior of casing in the pockets between the lobes and the casing. Finally, the meshing of the lobes forces liquid to pass through the outlet port. The bearings are placed out of the pumped liquid. Therefore the pressure is limited by the bearing location and shaft deflection.

Because of superb sanitary qualities, high efficiency, reliability, corrosion resistance and good clean-in-place and steam-in-place (CIP/SIP) characteristics, Lobe pumps are widely used in industries such as pulp and paper, chemical, food, beverage, pharmaceutical and biotechnology etc. These pumps can handle solids (e.g., cherries and olives), slurries, pastes, and a variety of liquids. A gentle pumping action minimizes product degradation.

They also offer continuous and intermittent reversible flows. Flow is relatively independent of changes in process pressure and therefore, the output is constant and continuous.

Lobe pumps are frequently used in food applications because they handle solids without damaging the product. Large sized particles can be pumped much effectively than in other positive displacement types. As the lobes do not make any direct contact therefore, the clearance is not as close as in other Positive displacement pumps. This specific design of pump makes it suitable to handle low viscosity fluids with diminished performance.

Loading characteristics are not as good as other designs, and suction ability is low. High viscosity liquids require reduced speeds to achieve satisfactory performance. The reduction in speed can be 25% or more in case of high viscosity fluid.

Internal Gear Pump

Internal gear pumps are exceptionally versatile. They are often used for low or medium viscosity fluids such as solvents and fuel oil and wide range of temperature. This is non-pulsing, self-priming and can run dry for short periods. It is a variation of the basic gear pump.

It comprises of an internal gear, a regular spur gear, a crescent-shaped seal and an external housing. The schematic of internal gear pump. Liquid enters the suction port between the rotor (large exterior gear) and idler (small interior gear) teeth. Liquid travels through the pump between the teeth and crescent. Crescent divides the liquid and acts as a seal between the suction and discharge ports. When the teeth mesh on the side opposite to the crescent seal, the fluid is forced out through the discharge port of the pump. This clearance between gears can be adjusted to accommodate high temperature, to handle high viscosity fluids and to accommodate the wear. These pumps are bi-rotational so that they can be used to load and unload the vessels. As these pumps have only two moving parts and one stuffing box, therefore they are reliable, simple to operate and easy to maintain. However, these pumps are not suitable for high speed and high pressure applications. Only one bearing is used in the pump therefore overhung load on shaft bearing reduces the life of the bearing.

Applications

Some common internal gear pump applications are:

- All varieties of fuel oil and lube oil

- Resins and Polymers

- Alcohols and solvents

- Asphalt, Bitumen, and Tar

- Polyurethane foam (Isocyanate and polyol)
- Food products such as corn syrup, chocolate, and peanut butter
- Paint, inks, and pigments
- Soaps and surfactants
- Glycol

Gerotor Pump

Gerotor is a positive displacement pump. The name Gerotor is derived from "Generated

Rotor". At the most basic level, a Gerotor is essentially one that is moved via fluid power.

Originally this fluid was water, today the wider use is in hydraulic devices. The schematic of Gerotor pump is shown in figure 5.2.5. Gerotor pump is an internal gear pump without the crescent. It consists of two rotors viz. inner and outer rotor. The inner rotor has N teeth, and the outer rotor has N+1 teeth. The inner rotor is located off-center and both rotors rotate. The geometry of the two rotors partitions the volume between them into N different dynamically-changing volumes. During the rotation, volume of each partition changes continuously. Therefore, any given volume first increases, and then decreases. An increase in volume creates vacuum. This vacuum creates suction, and thus, this part of the cycle sucks the fluid. As the volume decreases, compression occurs.

During this compression period, fluids can be pumped, or compressed (if they are gaseous fluids).

The close tolerance between the gears acts as a seal between the suction and discharge ports. Rotor and idler teeth mesh completely to form a seal equidistant from the discharge and suction ports. This seal forces the liquid out of the discharge port. The flow output is uniform and constant at the outlets.

The important advantages of the pumps are high speed operation, constant discharge in all pressure conditions, bidirectional operation, less sound in running condition and less maintenance due to only two moving parts and one stuffing box etc. However, the pump is having some limitations such as medium pressure operating range, clearance is fixed, solids can't be pumped and overhung load on the shaft bearing etc.

Applications

Gerotors are widely used in industries and are produced in variety of shapes and sizes by a number of different methods. These pumps are primarily suitable for low pressure applications such as lubrication systems or hot oil filtration systems, but can also be found in low to moderate pressure hydraulic applications. However common applications are as follows:

• Light fuel oils

• Lube oil

• Cooking oils

• Hydraulic fluid

This article is about power machinery. For civil engineering concerning water management, see Hydraulics.

"Hydraulic equipment" redirects here. For exercise equipment using hydraulic cylinders for resistance, see Resistance training.

A simple open center hydraulic circuit.

An excavator; main hydraulics: Boom cylinders, swing drive, cooler fan and track drive

Fundamental features of using hydraulics compared to mechanics for force and torque increase/decrease in a transmission.

This article has multiple issues. Please help improve it or discuss these issues on the talk page. (Learn how and when to remove these template messages)

This article needs additional citations for verification

This article provides insufficient context for those unfamiliar with the subject.

Hydraulic machines use liquid fluid power to perform work. Heavy construction vehicles are a common example. In this type of machine, hydraulic fluid is pumped to various hydraulic motors and hydraulic cylinders throughout the machine and becomes pressurized according to the resistance present. The fluid is controlled directly or automatically by control valves and distributed through hoses, tubes, and/or pipes.

Hydraulic systems, like pneumatic systems, are based on Pascal's law which states that any pressure applied to a fluid inside a closed system will transmit that pressure equally everywhere and in all directions. A hydraulic system uses an in-compressible liquid as its fluid, rather than a compressible gas.

The popularity of hydraulic machinery is due to the very large amount of power that can be transferred through small tubes and flexible hoses, and the high power density and wide array of actuators that can make use of this power, and the huge multiplication of forces that can be achieved by applying pressures over relatively large areas. One drawback, compared to machines using gears and shafts, is that any transmission of power results in some losses due to resistance of fluid flow through the piping.

Two hydraulic cylinders interconnected

Cylinder C1 is one inch in radius, and cylinder C2 is ten inches in radius. If the force exerted on C1 is 10 lbf, the force exerted by C2 is 1000 lbf because C2 is a hundred times larger in area ($S = \pi r^2$) as C1. The downside to this is that you have to move C1 a hundred inches to move C2 one inch. The most common use for this is the classical hydraulic jack where a pumping cylinder with a small diameter is connected to the lifting cylinder with a large diameter.

Pump and motor

If a hydraulic rotary pump with the displacement 10 cc/rev is connected to a hydraulic rotary motor with 100 cc/rev, the shaft torque required to drive the pump is one tenth of the torque then available at the motor shaft, but the shaft speed (rev/min) for the motor is also only one tenth of the pump shaft speed. This combination is actually the same type of force multiplication as the cylinder example, just that the linear force in this case is a rotary force, defined as torque.

Both these examples are usually referred to as a hydraulic transmission or hydrostatic transmission involving a certain hydraulic "gear ratio".

Hydraulic circuits

A hydraulic circuit is a system comprising an interconnected set of discrete components that transport liquid. The purpose of this system may be to control where fluid flows (as in a network of tubes of coolant in a thermodynamic system) or to control fluid pressure (as in hydraulic amplifiers). For example, hydraulic machinery uses hydraulic circuits (in which hydraulic fluid is pushed, under pressure, through hydraulic pumps, pipes, tubes, hoses, hydraulic motors, hydraulic cylinders, and so on) to move heavy loads. The approach of describing a fluid system in terms of discrete components is inspired by the success of electrical circuit theory. Just as electric circuit theory works when elements are discrete and linear, hydraulic circuit theory works best when the elements (passive component such as pipes or transmission lines or active components such as power packs or pumps) are discrete and linear. This usually means that hydraulic circuit analysis works best for long, thin tubes with discrete pumps, as found in chemical process flow systems or microscale devices.

The circuit comprises the following components:

- Active components
- Hydraulic power pack
- Transmission lines
- Hydraulic hoses
- Passive components
- Hydraulic cylinders

For the hydraulic fluid to do work, it must flow to the actuator and/or motors, then return to a reservoir. The fluid is then filtered and re-pumped. The path taken by hydraulic fluid is called a hydraulic circuit of which there are several types.

Open center circuits use pumps which supply a continuous flow. The flow is returned to tank through the control valve's open center; that is, when the control valve is centered, it provides an open return path to tank and the fluid is not pumped to a high pressure. Otherwise, if the control valve is actuated it routes fluid to and from an actuator and tank. The fluid's pressure will rise to meet any resistance, since the pump has a constant output. If the pressure rises too high, fluid returns to tank through a pressure relief valve. Multiple control valves may be stacked in series. This type of circuit can use inexpensive, constant displacement pumps.

Closed center circuits supply full pressure to the control valves, whether any valves are actuated or not. The pumps vary their flow rate, pumping very little hydraulic fluid until the operator actuates a valve. The valve's spool therefore doesn't need an open center return path to tank. Multiple valves can be connected in a parallel arrangement and system pressure is equal for all valves.

Open loop and closed loop circuits

Open loop circuits

Open-loop: Pump-inlet and motor-return (via the directional valve) are connected to the hydraulic tank. The term loop applies to feedback; the more correct term is open versus closed "circuit". Open center circuits use pumps which supply a continuous flow. The flow is returned to the tank through the control valve's open center; that is, when the control valve is centered, it provides an open return path to the tank and the fluid is not pumped to a high pressure. Otherwise, if the control valve is actuated it routes fluid to and from an actuator and tank. The fluid's pressure will rise to meet any resistance, since the pump has a constant output. If the pressure rises too high, fluid returns to the tank through a pressure relief valve. Multiple control valves may be stacked in series. This type of circuit can use inexpensive, constant displacement pumps.

Closed loop circuits

Closed-loop: Motor-return is connected directly to the pump-inlet. To keep up pressure on the low pressure side, the circuits have a charge pump (a small gear pump) that supplies cooled and filtered oil to the low pressure side. Closed-loop circuits are generally used for hydrostatic transmissions in mobile applications. Advantages: No directional valve and better response, the circuit can work with higher pressure. The pump swivel angle covers both positive and negative flow direction. Disadvantages: The pump cannot be utilized for any other hydraulic function in an easy way and cooling can be a problem due to limited exchange of oil flow. High power closed loop systems generally must have a 'flush-valve' assembled in the circuit in order to exchange much more flow than the basic leakage flow from the pump and the motor, for increased cooling and filtering. The flush valve is normally integrated in the motor housing to get a cooling effect for the oil that is rotating in the motor housing itself. The losses in the motor housing from rotating effects and losses in the ball bearings can be considerable as motor speeds will reach 4000-5000 rev/min or even more at maximum vehicle speed. The leakage flow as well as the extra flush flow must be supplied by the charge pump. A large charge pump is thus very important if the transmission is designed for high pressures and high motor speeds. High oil temperature is usually a major problem when using hydrostatic transmissions at high vehicle speeds for longer periods, for instance when transporting the machine from one work place to the other. High oil temperatures for long periods will drastically reduce the lifetime of the transmission. To keep down the oil temperature, the system pressure during transport must be lowered, meaning that the minimum displacement for the motor must be limited to a reasonable value. Circuit pressure during transport around 200-250 bar is recommended.

Closed loop systems in mobile equipment are generally used for the transmission as an alternative to mechanical and hydrodynamic (converter) transmissions. The advantage is a stepless gear ratio (continuously variable speed/torque) and a more flexible control of the gear ratio depending on the load and operating conditions. The hydrostatic transmission is generally limited to around 200 kW maximum power, as the total cost gets too high at higher power compared to a hydrodynamic transmission. Large wheel loaders for instance and heavy machines are therefore usually equipped with converter transmissions. Recent technical achievements for the converter transmissions have improved the efficiency and developments in the software have also improved the characteristics, for example selectable gear shifting programs during operation and more gear steps, giving them characteristics close to the hydrostatic transmission.

Constant pressure and load-sensing systems

Hydrostatic transmissions for earth moving machines, such as for track loaders, are often equipped with a separate 'inch pedal' that is used to temporarily increase the diesel engine rpm while reducing the vehicle speed in order to increase the available hydraulic power output for the working hydraulics at low speeds and increase the tractive effort. The function is similar to stalling a converter gearbox at high engine rpm. The inch function affects the preset characteristics for the 'hydrostatic' gear ratio versus diesel engine rpm.

Constant pressure (CP) systems

The closed center circuits exist in two basic configurations, normally related to the regulator for the variable pump that supplies the oil:

Constant pressure systems (CP-system), standard. Pump pressure always equals the pressure setting for the pump regulator. This setting must cover the maximum required load pressure. Pump delivers flow according to required sum of flow to the consumers. The CP-system generates large power losses if the machine works with large variations in load pressure and the average system pressure is much lower than the pressure setting for the pump regulator. CP is simple in design, and works like a pneumatic system. New hydraulic functions can easily be added and the system is quick in response.

Constant pressure systems (CP-system), unloaded. Same basic configuration as 'standard' CP-system but the pump is unloaded to a low stand-by pressure when all valves are in neutral position.

Not so fast response as standard CP but pump lifetime is prolonged.

Load-sensing (LS) systems

Load-sensing systems (LS-system) generates less power losses as the pump can reduce both flow and pressure to match the load requirements, but requires more tuning than the CP-system with respect to system stability. The LS-system also requires additional logical valves and compensator valves in the directional valves, thus it is technically more complex and more expensive than the CP-system. The LS-system generates a constant power loss related to the regulating pressure drop for the pump regulator:

 If the pump flow is high the extra loss can be considerable. The power loss also increases if the load pressures vary a lot. The cylinder areas, motor displacements and mechanical torque arms must be designed to match load pressure in order to bring down the power losses. Pump pressure always equals the maximum load pressure when several functions are run simultaneously and the power input to the pump equals the (max. load pressure + ΔpLS) x sum of flow.

Five basic types of load-sensing systems

Load sensing without compensators in the directional valves. Hydraulically controlled LS-pump.

Load sensing with up-stream compensator for each connected directional valve. Hydraulically controlled LS-pump.

Load sensing with down-stream compensator for each connected directional valve. Hydraulically controlled LS-pump.

Load sensing with a combination of up-stream and down-stream compensators. Hydraulically controlled LS-pump.

Load sensing with synchronized, both electric controlled pump displacement and electric controlled

valve flow area for faster response, increased stability and fewer system losses. This is a new type of LS-system, not yet fully developed.

Technically the down-stream mounted compensator in a valveblock can physically be mounted "up-stream", but work as a down-stream compensator.

System type gives the advantage that activated functions are synchronized independent of pump flow capacity. The flow relation between 2 or more activated functions remains independent of load pressures, even if the pump reaches the maximum swivel angle. This feature is important for machines that often run with the pump at maximum swivel angle and with several activated functions that must be synchronized in speed, such as with excavators. With typ system, the functions with up-stream compensators have priority. Example: Steering-function for a wheel loader. The system type with down-stream compensators usually have a unique trademark depending on the manufacturer of the valves, for example "LSC" (Linde Hydraulics), "LUDV" (Bosch Rexroth Hydraulics) and "Flowsharing" (Parker Hydraulics) etc. No official standardized name for this type of system has been established but Flowsharing is a common name for it.

Hydraulic pump

An exploded view of an external gear pump.

Hydraulic pumps supply fluid to the components in the system. Pressure in the system develops in reaction to the load. Hence, a pump rated for 5,000 psi is capable of maintaining flow against a load of 5,000 psi.

Pumps have a power density about ten times greater than an electric motor (by volume). They are powered by an electric motor or an engine, connected through gears, belts, or a flexible elastomeric coupling to reduce vibration.

Common types of hydraulic pumps to hydraulic machinery applications are;

1. Gear pump: cheap, durable (especially in g-rotor form), simple. Less efficient, because they are constant (fixed) displacement, and mainly suitable for pressures below 20 MPa (3000 psi).
2. Vane pump: cheap and simple, reliable. Good for higher-flow low-pressure output.
3. Axial piston pump: many designed with a variable displacement mechanism, to vary output flow for automatic control of pressure. There are various axial piston pump designs, including swashplate (sometimes referred to as a valveplate pump) and checkball (sometimes referred to as a wobble plate pump). The most common is the swashplate pump. A variable-angle swashplate causes the pistons to reciprocate a greater or lesser distance per rotation, allowing output flow rate

and pressure to be varied (greater displacement angle causes higher flow rate, lower pressure, and vice versa).
4. Radial piston pump: normally used for very high pressure at small flows.

Piston pumps are more expensive than gear or vane pumps, but provide longer life operating at higher pressure, with difficult fluids and longer continuous duty cycles. Piston pumps make up one half of a hydrostatic transmission.

Control valves

Control valves on a scissor lift

Directional control valves route the fluid to the desired actuator. They usually consist of a spool inside a cast iron or steel housing. The spool slides to different positions in the housing, and intersecting grooves and channels route the fluid based on the spool's position.

The spool has a central (neutral) position maintained with springs; in this position the supply fluid is blocked, or returned to tank. Sliding the spool to one side routes the hydraulic fluid to an actuator and provides a return path from the actuator to tank. When the spool is moved to the opposite direction the supply and return paths are switched. When the spool is allowed to return to neutral (center) position the actuator fluid paths are blocked, locking it in position.

Directional control valves are usually designed to be stackable, with one valve for each hydraulic cylinder, and one fluid input supplying all the valves in the stack.

Tolerances are very tight in order to handle the high pressure and avoid leaking, spools typically have a clearance with the housing of less than a thousandth of an inch (25 μm). The valve block will be mounted to the machine's frame with a three point pattern to avoid distorting the valve block and jamming the valve's sensitive components.

The spool position may be actuated by mechanical levers, hydraulic pilot pressure, or solenoids which push the spool left or right. A seal allows part of the spool to protrude outside the housing, where it is accessible to the actuator.

The main valve block is usually a stack of off the shelf directional control valves chosen by flow capacity and performance. Some valves are designed to be proportional (flow rate proportional to valve position), while others may be simply on-off. The control valve is one of the most expensive and sensitive parts of a hydraulic circuit.

Pressure relief valves are used in several places in hydraulic machinery; on the return circuit to maintain a small amount of pressure for brakes, pilot lines, etc... On hydraulic cylinders, to prevent overloading and hydraulic line/seal rupture. On the hydraulic reservoir, to maintain a small positive pressure which excludes moisture and contamination.

Pressure regulators reduce the supply pressure of hydraulic fluids as needed for various circuits.

Sequence valves control the sequence of hydraulic circuits; to ensure that one hydraulic cylinder is fully extended before another starts its stroke, for example.

Shuttle valves provide a logical or function.

Check valves are one-way valves, allowing an accumulator to charge and maintain its pressure after the machine is turned off, for example.

Pilot controlled check valves are one-way valve that can be opened (for both directions) by a foreign pressure signal. For instance if the load should not be held by the check valve anymore. Often the foreign pressure comes from the other pipe that is connected to the motor or cylinder.

Counterbalance valves are in fact a special type of pilot controlled check valve. Whereas the check valve is open or closed, the counterbalance valve acts a bit like a pilot controlled flow control.

Cartridge valves are in fact the inner part of a check valve; they are off the shelf components with a standardized envelope, making them easy to populate a proprietary valve block. They are available in many configurations; on/off, proportional, pressure relief, etc. They generally screw into a valve block and are electrically controlled to provide logic and automated functions.

Hydraulic fuses are in-line safety devices designed to automatically seal off a hydraulic line if pressure becomes too low, or safely vent fluid if pressure becomes too high.

Auxiliary valves in complex hydraulic systems may have auxiliary valve blocks to handle various duties unseen to the operator, such as accumulator charging, cooling fan operation, air conditioning power, etc. They are usually custom valves designed for the particular machine, and may consist of a metal block with ports and channels drilled. Cartridge valves are threaded into the ports and may be electrically controlled by switches or a microprocessor to route fluid power as needed.

Actuators

Hydraulic cylinder

Hydraulic motor (a pump plumbed in reverse); hydraulic motors with axial configuration use Swashplates for highly accurate control and also in 'no stop' continuous (360°) precision positioning mechanisms. These are frequently driven by several hydraulic pistons acting in sequence.

Hydrostatic transmission

Brakes

Reservoir

The hydraulic fluid reservoir holds excess hydraulic fluid to accommodate volume changes from: cylinder extension and contraction, temperature driven expansion and contraction, and leaks. The reservoir is also designed to aid in separation of air from the fluid and also work as a heat accumulator to cover losses in the system when peak power is used. Design engineers are always pressured to reduce the size of hydraulic reservoirs, while equipment operators always appreciate larger reservoirs.

Reservoirs can also help separate dirt and other particulate from the oil, as the particulate will generally settle to the bottom of the tank. Some designs include dynamic flow channels on the fluid's return path that allow for a smaller reservoir.

Accumulators

Accumulators are a common part of hydraulic machinery. Their function is to store energy by using pressurized gas. One type is a tube with a floating piston. On the one side of the piston there is a charge of pressurized gas, and on the other side is the fluid. Bladders are used in other designs. Reservoirs store a system's fluid.

Examples of accumulator uses are backup power for steering or brakes, or to act as a shock absorber for the hydraulic circuit.,

Hydraulic fluid

Also known as tractor fluid, hydraulic fluid is the life of the hydraulic circuit. It is usually petroleum oil with various additives. Some hydraulic machines require fire resistant fluids, depending on their applications. In some factories where food is prepared, either an edible oil or water is used as a working fluid for health and safety reasons.

In addition to transferring energy, hydraulic fluid needs to lubricate components, suspend contaminants and metal filings for transport to the filter, and to function well to several hundred degrees Fahrenheit or Celsius.

Filters

Filters are an important part of hydraulic systems which removes the unwanted particles from fluid. Metal particles are continually produced by mechanical components and need to be removed along with other contaminants.

Filters may be positioned in many locations. The filter may be located between the reservoir and the pump intake. Blockage of the filter will cause cavitation and possibly failure of the pump. Sometimes the filter is located between the pump and the control valves. This arrangement is more expensive, since the filter housing is pressurized, but eliminates cavitation problems and protects the control valve from pump failures. The third common filter location is just before the return line enters the reservoir. This location is relatively insensitive to blockage and does not require a pressurized housing, but contaminants that enter the reservoir from external sources are not filtered until passing through the system at least once. Filters are used from 7 micron to 15 micron depends upon the viscosity grade of hydraulic oil.

Tubes, pipes and hoses

Hydraulic tubes are seamless steel precision pipes, specially manufactured for hydraulics. The tubes have standard sizes for different pressure ranges, with standard diameters up to 100 mm. The tubes are supplied by manufacturers in lengths of 6 m, cleaned, oiled and plugged. The tubes are interconnected by different types of flanges (especially for the larger sizes and pressures), welding cones/nipples (with o-ring seal), several types of flare connection and by cut-rings. In larger sizes, hydraulic pipes are used. Direct joining of tubes by welding is not acceptable since the interior cannot be inspected.

Hydraulic pipe is used in case standard hydraulic tubes are not available. Generally these are used for low pressure. They can be connected by threaded connections, but usually by welds. Because of the larger diameters the pipe can usually be inspected internally after welding. Black pipe is non-galvanized and suitable for welding.

Hydraulic hose is graded by pressure, temperature, and fluid compatibility. Hoses are used when pipes or tubes can not be used, usually to provide flexibility for machine operation or maintenance. The hose is built up with rubber and steel layers. A rubber interior is surrounded by multiple layers of woven wire and rubber. The exterior is designed for abrasion resistance. The bend radius of hydraulic hose is carefully designed into the machine, since hose failures can be deadly, and violating the hose's minimum bend radius will cause failure. Hydraulic hoses generally have steel fittings swaged on the ends. The weakest part of the high pressure hose is the connection of the hose to the fitting. Another disadvantage of hoses is the shorter life of rubber which requires periodic replacement, usually at five to seven year intervals.

Tubes and pipes for hydraulic n applications are internally oiled before the system is commissioned. Usually steel piping is painted outside. Where flare and other couplings are used, the paint is removed under the nut, and is a location where corrosion can begin. For this reason, in marine applications most piping is stainless steel.

Seals, fittings and connections

Main article: Seal (mechanical)

Components of a hydraulic system [sources (e.g. pumps), controls (e.g. valves) and actuators (e.g. cylinders)] need connections that will contain and direct the hydraulic fluid without leaking or losing the pressure that makes them work. In some cases, the components can be made to bolt together with fluid paths built-in. In more cases, though, rigid tubing or flexible hoses are used to direct the flow from one component to the next. Each component has entry and exit points for the fluid involved (called ports) sized according to how much fluid is expected to pass through it.

There are a number of standardized methods in use to attach the hose or tube to the component. Some are intended for ease of use and service, others are better for higher system pressures or control of leakage. The most common method, in general, is to provide in each component a female-threaded port, on each hose or tube a female-threaded captive nut, and use a separate adapter fitting with matching male threads to connect the two. This is functional, economical to manufacture, and easy to service.

Fittings serve several purposes;

To join components with ports of different sizes.

To bridge different standards; O-ring boss to JIC, or pipe threads to face seal, for example.

To allow proper orientation of components, a 90°, 45°, straight, or swivel fitting is chosen as needed. They are designed to be positioned in the correct orientation and then tightened.

To incorporate bulkhead hardware to pass the fluid through an obstructing wall.

A quick disconnect fitting may be added to a machine without modification of hoses or valves

A typical piece of machinery or heavy equipment may have thousands of sealed connection points and several different types:

Pipe fittings, the fitting is screwed in until tight, difficult to orient an angled fitting correctly without over or under tightening.

O-ring boss, the fitting is screwed into a boss and orientated as needed, an additional nut tightens the fitting, washer and o-ring in place.

Flare fittings, are metal to metal compression seals deformed with a cone nut and pressed into a flare mating.

Face seal, metal flanges with a groove and o-ring seal are fastened together.

Beam seals are costly metal to metal seals used primarily in aircraft.

Swaged seals, tubes are connected with fittings that are swaged permanently in place. Primarily used in aircraft.

Elastomeric seals (O-ring boss and face seal) are the most common types of seals in heavy equipment and are capable of reliably sealing 6000+ psi (40+ MPa) of fluid pressure.

What Is a Hydraulic System?

Hydraulic System

With a variety of applications, hydraulic systems are used in all kinds of large and small industrial settings, as well as buildings, construction equipment, and vehicles. Paper mills, logging, manufacturing, robotics, and steel processing are leading users of hydraulic equipment.

As an efficient and cost-effective way to create movement or repetition, hydraulic system-based equipment is hard to top. It's likely your company has hydraulics in use in one or more applications for these reasons.

We'll provide more information about hydraulic systems in this article, including covering the definition and basic designs and components.

Convergence Training are manufacturing and industrial maintenance training experts. Click the links below to learn more about how we can help you.

An Overview of Hydraulic Systems

The purpose of a specific hydraulic system may vary, but all hydraulic systems work through the same basic concept. Defined simply, hydraulic systems function and perform tasks through using a fluid that is pressurized. Another way to put this is the pressurized fluid makes things work.

The power of liquid fuel in hydraulics is significant and as a result, hydraulic are commonly used in heavy equipment. In a hydraulic system, pressure, applied to a contained fluid at any point, is transmitted undiminished. That pressurized fluid acts upon every part of the section of a containing vessel and creates force or power. Due to the use of this force, and depending on how it's applied, operators can lift heavy loads, and precise repetitive tasks can be easily done.

Marvelously versatile, hydraulic systems are dynamic, yet relatively straightforward in how they work.

Let's look at some applications and a few basic components found in hydraulic systems. This short sample from our online hydraulic systems and components course sets the scene nicely.

Hydraulic Circuits

Transporting liquid through a set of interconnected discrete components, a hydraulic circuit is a system that can control where fluid flows (such as thermodynamic systems), as well as control fluid pressure (such as hydraulic amplifiers).

The system of a hydraulic circuit works similar to electric circuit theory, using linear and discrete elements. Hydraulic circuits are often applied in chemical processing (flow systems).

Hydraulic Pumps

Mechanical power is converted into hydraulic energy using the flow and pressure of a hydraulic pump. Hydraulic pumps operate by creating a vacuum at a pump inlet, forcing liquid from a reservoir into an inlet line, and to the pump. Mechanical action sends the liquid to the pump outlet, and as it does, forces it into the hydraulic system.

This is an example of Pascal's Law, which is foundational to the principle of hydraulics. According to Pascal's Law, "A pressure change occurring anywhere in a confined incompressible fluid is transmitted throughout the fluid such that the same change occurs everywhere."

Hydraulic Motors

The conversion of hydraulic pressure and flow into torque (or a twisting force) and then rotation is the function of a hydraulic motor, which is a mechanical actuator.

The use of these is quite adaptable. Along with hydraulic cylinders and hydraulic pumps, hydraulic motors can be united in a hydraulic drive system. Combined with hydraulic pumps, the hydraulic motors can create hydraulic transmissions. While some hydraulic motors run on water, the majority in today's business operations are powered by hydraulic fluid, as the ones in your business likely are.

Hydraulic Cylinders

A hydraulic cylinder is a mechanism that converts energy stored in the hydraulic fluid into a force used to move the cylinder in a linear direction. It too has many applications and can be either single acting or double acting. As part of the complete hydraulic system, the cylinders initiate the pressure of the fluid, the flow of which is regulated by a hydraulic motor.

Hydraulic Energy and Safety

Hydraulics present a set of hazards to be aware of, and for that reason safety training is required.

For example, this short sample from our online hydraulic safety training course explains some of the ways the fluids in a hydraulic system can be hazardous.

Remember, the purpose of hydraulic systems is to create motion or force. It's a power source, generating energy.

Don't underestimate hydraulic energy in your safety program. It is small but mighty in force. And like any force, it can do great good or great harm.

In the workplace, that translates to a potential hazard source, especially if uncontrolled. Hydraulic energy is subject to OSHA's Lockout/Tagout rules, along with electrical energy and other similar hazard sources. Be sure to train workers about the hazards of uncontrolled hydraulic energy, especially during maintenance, and the need for lockout/tagout, as illustrated by this still image from one of our online lockout/tagout training courses.

If neglected in procedures or forgotten when servicing equipment, uncontrolled hydraulic energy can have devastating results. Failure to control hydraulic energy frequently causes crushing events, amputations, and lacerations to exposed workers.

Therefore, like other energy sources, hydraulic energy must be controlled, using an appropriate energy isolating device that prevents a physical release of energy. There are also systems that require the release of stored hydraulic energy to relieve pressure. And also, those engaged in lockout/tagout, must also verify the release of stored hydraulic energy/pressure (usually indicated by zero pressure on gauges) prior to working on equipment.

Also, workers need training which must explain the hazard potential and clearly detail methods to prevent injury. According to OSHA:

"All employees who are authorized to lockout machines or equipment and perform the service and maintenance operations need to be trained in recognition of applicable hazardous energy sources in the workplace, the type and magnitude of energy found in the workplace, and the means and methods of isolating and/or controlling the energy."

You should be very familiar with any equipment in your business that creates hydraulic energy to ensure your workers are adequately protected through well-detailed procedures and training. And of course, your LO/TO program should echo your procedures, and list sources of workplace hydraulic energy devices. (Don't forget to perform at least annual reviews of the program and procedures to ensure you catch any changes or deficiencies.)

Again, it's critical anyone involved with hydraulic systems is properly trained. Don't neglect that aspect.

If you'd like to dig deeper into this topic, we have several courses on hydraulic systems, including Hydraulic System Basics, which outlines the essentials of hydraulic theory, common components, what mechanical advantage is, and how hydraulic fluid is contaminated. In addition, we have two others which provide vital in-depth information, Hydraulic System Valves and Components and Hydraulic System Equipment.

You might also want to check out our series of courses including materials on hydraulics.

It's important to understand the principles of these systems, not only for servicing and maintenance, but also to understand the ways the hydraulic systems function to avoid injuries and accidents.

What is a hydraulic system?

A hydraulic system is a drive technology where a fluid is used to move the energy from e.g. an electric motor to an actuator, such as a hydraulic cylinder. The fluid is theoretically uncompressible and the fluid path can be flexible in the same way as an electric cable.

What is a hydraulic system used for?

Hydraulic systems are mainly used where a high power density is needed or load requirements chance rapidly. This is especially the case in all kinds of mobile equipment such as excavators and in industrial systems such as presses.

In wind turbines, hydraulics is used for pitch and brake control. In some cases, different auxiliary systems such as hatches and cranes are also powered by hydraulic systems.

Why are hydraulic systems used?

The main reason for using hydraulics is the high power density and secondly the simplicity coming from using few components to realize complex and fast moving machines with a high degree of safety.

Hydraulic Systems and Fluid Selection

It wasn't until the beginning of the industrial revolution when a British mechanic applied the principle of Pascal's law in the development of the first hydraulic press. In 1795, he patented his hydraulic press, known as the Bramah press. Bramah figured that if a small force on a small area would create a proportionally larger force on a larger area, the only limit to the force that a machine can exert is the area to which the pressure is applied.

What is a Hydraulic System?

Hydraulic systems can be found today in a wide variety of applications, from small assembly processes to integrated steel and paper mill applications. Hydraulics enable the operator to accomplish significant work (lifting heavy loads, turning a shaft, drilling precision holes, etc.) with a minimum investment in mechanical linkage through the application of Pascal's law, which states:

"Pressure applied to a confined fluid at any point is transmitted undiminished throughout the fluid in all directions and acts upon every part of the confining vessel at right angles to its interior surfaces and equally upon equal areas."

By applying Pascal's law and Brahma's application of it, it is evident that an input force of 100 pounds on 10 square inches will develop a pressure of 10 pounds per square inch throughout the confined vessel. This pressure will support a 1000-pound weight if the area of the weight is 100 square inches.

The principle of Pascal's law is realized in a hydraulic system by the hydraulic fluid that is used to transmit the energy from one point to another. Because hydraulic fluid is nearly incompressible, it is able to transmit power instantaneously.

Hydraulic System Components

The major components that make up a hydraulic system are the reservoir, pump, valve(s) and actuator(s) (motor, cylinder, etc.).

Reservoir

The purpose of the hydraulic reservoir is to hold a volume of fluid, transfer heat from the system, allow solid contaminants to settle and facilitate the release of air and moisture from the fluid.

Pump

The hydraulic pump transmits mechanical energy into hydraulic energy. This is done by the movement of fluid which is the transmission medium. There are several types of hydraulic pumps including gear, vane and piston. All of these pumps have different subtypes intended for specific applications such as a bent-axis piston pump or a variable displacement vane pump. All hydraulic pumps work on the same principle, which is to displace fluid volume against a resistant load or pressure.

Valves

Hydraulic valves are used in a system to start, stop and direct fluid flow. Hydraulic valves are made up of poppets or spools and can be actuated by means of pneumatic, hydraulic, electrical, manual or mechanical means.

Actuators

Hydraulic actuators are the end result of Pascal's law. This is where the hydraulic energy is converted back to mechanical energy. This can be done through use of a hydraulic cylinder which converts hydraulic energy into linear motion and work, or a hydraulic motor which converts hydraulic energy into rotary motion and work. As with hydraulic pumps, hydraulic cylinders and hydraulic motors have several different subtypes, each intended for specific design applications.

Key Lubricated Hydraulic Components

There are several components in a hydraulic system, that due to cost of repair or criticality of mission, are considered vital components. Pumps and valves are considered key components. Several different configurations for pumps must be treated individually from a lubrication perspective, including:

Vane Pumps

There are many variations of vane pumps available between manufacturers. They all work on similar design principles. A slotted rotor is coupled to the drive shaft and turns inside of a cam ring that is offset or eccentric to the drive shaft. Vanes are inserted into the rotor slots and follow the inner surface of the cam ring as the rotor turns.

The vanes and the inner surface of the cam rings are always in contact and are subject to high amounts of wear. As the two surfaces wear, the vanes come further out of their slot. Vane pumps deliver a steady flow at a high cost. Vane pumps operate at a normal viscosity range between 14 and 160 cSt at operating temperature. Vane pumps may not be suitable in critical high-pressure hydraulic systems where contamination and fluid quality are difficult to control. The performance of the fluid's antiwear additive is generally very important with vane pumps.

Piston Pumps

As with all hydraulic pumps, piston pumps are available in fixed and variable displacement designs. Piston pumps are generally the most versatile and rugged pump type and offer a range of options for any type of system. Piston pumps can operate at pressures beyond 6000 psi, are highly efficient and produce comparatively little noise. Many designs of piston pumps also tend to resist wear better than other pump types. Piston pumps operate at a normal fluid viscosity range of 10 to 160 cSt.

Gear Pumps

There are two common types of gear pumps, internal and external. Each type has a variety of subtypes, but all of them develop flow by carrying fluid between the teeth of a meshing gear set. While generally less efficient than vane and piston pumps, gear pumps are often more tolerant of fluid contamination.

Internal gear pumps produce pressures up to 3000 to 3500 psi. These types of pumps offer a wide viscosity range up to 2200 cSt, depending on flow rate and are generally quiet. Internal gear pumps also have a high efficiency even at low fluid viscosity.

External gear pumps are common and can handle pressures up to 3000 to 3500 psi. These gear pumps offer an inexpensive, mid-pressure, mid-volume, fixed isplacement delivery to a system.

Viscosity ranges for these types of pumps are limited to less than 300 cSt.

Hydraulic Fluids

Today's hydraulic fluids serve multiple purposes. The major function of a hydraulic fluid is to provide energy transmission through the system which enables work and motion to be accomplished. Hydraulic fluids are also responsible for lubrication, heat transfer and contamination control. When selecting a lubricant, consider the viscosity, seal compatibility, basestock and the additive package. Three common varieties of hydraulic fluids found on the market today are petroleum-based, water-based and synthetics.

Petroleum-based or mineral-based fluids are the most widely used fluids today. The properties of a mineral-based fluid depend on the additives used, the quality of the original crude oil and the refining process. Additives in a mineral-based fluid offer a range of specific performance characteristics.

Common hydraulic fluid additives include rust and oxidation inhibitors (R&O), anticorrosion agents, demulsifiers, antiwear (AW) and extreme pressure (EP) agents, VI improvers and defoamants. Mineral-based fluids offer a low-cost, high quality, readily available selection.

Water-based fluids are used for fire-resistance due to their high-water content. They are available as oil-in-water emulsions, water-in-oil (invert) emulsions and water glycol blends. Water-based fluids can provide suitable lubrication characteristics but need to be monitored closely to avoid problems. Because water-based fluids are used in applications when fire resistance is needed, these systems and the atmosphere around the systems can be hot.

Elevated temperatures cause the water in the fluids to evaporate, which causes the viscosity to rise. Occasionally, distilled water will have to be added to the system to correct the balance of the fluid. Whenever these fluids are used, several system components must be checked for compatibility, including pumps, filters, plumbing, fittings and seal materials.

Water-based fluids can be more expensive than conventional petroleum-based fluids and have other disadvantages (for example, lower wear resistance) that must be weighed against the advantage of fire-resistance.

Synthetic fluids are man-made lubricants and many offer excellent lubrication characteristics in high-pressure and high-temperature systems. Some of the advantages of synthetic fluids may include fire-resistance (phosphate esters), lower friction, natural detergency (organic esters and ester-enhanced synthesized hydrocarbon fluids) and thermal stability.

The disadvantage to these types of fluids is that they are usually more expensive than conventional fluids, they may be slightly toxic and require special disposal, and they are often not compatible with standard seal materials.

Fluid Properties

When choosing a hydraulic fluid, consider the following characteristics: viscosity, viscosity index, oxidation stability and wear resistance. These characteristics will determine how your fluid operates within your system. Fluid property testing is done in accordance with either American Society of Testing and Materials (ASTM) or other recognized standards organizations.

Viscosity (ASTM D445-97) is the measure of a fluid's resistance to flow and shear. A fluid of higher viscosity will flow with higher resistance compared to a fluid with a low viscosity. Excessively high viscosity can contribute to high fluid temperature and greater energy consumption. Viscosity that is too high or too low can damage a system, and consequently, is the key factor when considering a hydraulic fluid.

Viscosity Index (ASTM D2270) is how the viscosity of a fluid changes with a change in temperature. A high VI fluid will maintain its viscosity over a broader temperature range than a low VI fluid of the same weight. High VI fluids are used where temperature extremes are expected. This is particularly important for hydraulic systems that operate outdoors.

Oxidation Stability (ASTM D2272 and others) is the fluid's resistance to heat-induced degradation caused by a chemical reaction with oxygen. Oxidation greatly reduces the life of a fluid, leaving by-products such as sludge and varnish. Varnish interferes with valve functioning and can restrict flow passageways.

Wear Resistance (ASTM D2266 and others) is the lubricant's ability to reduce the wear rate in frictional boundary contacts. This is achieved when the fluid forms a protective film on metal surfaces to prevent abrasion, scuffing and contact fatigue on component surfaces.

Ten Steps to Check Optimum Viscosity Range

When selecting lubricants, ensure that the lubricant performs efficiently at the operating parameters of the system pump or motor. It is useful to have a defined procedure to follow through the process. Consider a simple system with a fixed-displacement gear pump that drives a cylinder (Figure 2).

Collect all relevant data for the pump. This includes collecting all the design limitations and optimum operating characteristics from the manufacturer. What you are looking for is the optimum operating viscosity range for the pump in question. Minimum viscosity is 13 cSt, maximum viscosity is 54 cSt, and optimum viscosity is 23 cSt.

Check the actual operating temperature conditions of the pump during normal operation. This step is extremely important because it gives a reference point for comparing different fluids during operation. Pump normally operates at 92ºC.

Collect the temperature-viscosity characteristics of the lubricant in use. The ISO viscosity rating system (cSt at 40ºC and 100ºC) is recommended. Viscosity is 32 cSt at 40ºC and 5.1 cSt at 100ºC.

Obtain an ASTM D341 standard viscosity-temperature chart for liquid petroleum products. This chart is quite common and can be found in most industrial lubricant product guides (Figure 3) or from lubricant suppliers.

Using the viscosity characteristics of the lubricant found in Step 3, start at the temperature axis (x-axis) of the chart and scroll along until you find the 40-degree C line. At the 40-degree C line, track upward until you find the line corresponding to the viscosity of your lubricant at 40ºC as published by your lubricant manufacturer. When you find the corresponding line, make a small mark at the intersection of the two lines.

Repeat Step 5 for the lubricant properties at 100ºC and mark the intersection point.

Connect the marks by drawing a line through them with a straight edge (yellow line,). This line represents the lubricant's viscosity at a range of temperatures.

Using the manufacturer's data for the pump's optimum operating viscosity, find the value on the vertical viscosity axis of the chart. Draw a horizontal line across the page until it hits the yellow viscosity vs. temperature line of the lubricant. Now draw a vertical line (green line, Figure 5) to the bottom of the chart from the yellow viscosity vs. temperature line where it is intersected by the horizontal optimum viscosity line. Where this line crosses, the temperature axis is the optimum operating temperature of the pump for this specific lubricant (69ºC).

Repeat Step 8 for maximum continuous and minimum continuous viscosities of the pump (brown lines, Figure 5). The area between the minimum and maximum temperatures is the minimum and maximum allowable operating temperature of the pump for the selected lubricant product.

Find the normal operating temperature of the pump on the chart using the heat gun scan done in Step 2. If the value is within the minimum and maximum temperatures as outlined on the chart, the fluid is suitable for use in the system. If it is not, you must change the fluid to a higher or lower viscosity grade accordingly. As shown in the chart, the normal operating conditions of the pump are out of the suitable range (brown area,) for our particular lubricant and will have to be changed.

Consolidating Hydraulic Fluids

The purpose of hydraulic fluid consolidation is to reduce complexity and inventory. Caution must be observed to consider all of the critical fluid characteristics required for each system. Therefore, fluid consolidation needs to start at the system level. Consider the following when consolidating fluids:

Determine the specific requirements of each piece of equipment. Consider all the normal operating limits of your equipment.

Talk to your preferred lubricant representative. You can gather and relay important information about the lubrication needs of your equipment. This will ensure that your supplier has all the products you require. Don't sacrifice system requirements to achieve consolidation.

Also, observe the following hydraulic fluid management practices.

Implement a procedure for labeling all incoming lubricants and tagging all reservoirs. This will minimize cross-contamination and assure that critical performance requirements are met.

Use a First-In-First-Out (FIFO) method in your lubricant storage facility. A properly executed FIFO system reduces confusion and storage-induced lubricant failure.

Hydraulic systems are complicated fluid-based systems for transferring energy and converting that energy into useful work. Successful hydraulic operations require the careful selection of hydraulic fluids that meet the system demands. Viscosity selection is central to a correct fluid selection.

There are other important parameters to consider as well, including viscosity index, wear resistance and oxidation resistance. Fluids can often be consolidated to reduce complexity and material storage cost. Caution should be exercised to avoid sacrificing fluid performance in an effort to achieve fluid consolidation.

How Does a Hydraulic System Work?

Hydraulic systems can be found in everything from cars to industrial machinery. They're designed to provide power, control, safety, and reliability, but how does a hydraulic system work?

How Does a Hydraulic System Work?

How Does a Hydraulic System Work Hydraulic systems are made up of numerous parts:

- The reservoir holds hydraulic fluid.

- The hydraulic pump moves the liquid through the system and converts mechanical energy and motion into hydraulic fluid power.
- The electric motor powers the hydraulic pump.
- The valves control the flow of the liquid and relieve excessive pressure from the system if needed.
- The hydraulic cylinder converts the hydraulic energy back into mechanical energy.
- There are also numerous types of hydraulic systems, but each contains the same main components listed above. They're also all designed to work the same way.

Hydraulic systems use the pump to push hydraulic fluid through the system to create fluid power. The fluid passes through the valves and flows to the cylinder where the hydraulic energy converts back into mechanical energy. The valves help to direct the flow of the liquid and relieve pressure when needed.

Hydraulic Systems on Ships

How Does a Hydraulic System Work In addition to vehicles and industrial machinery, hydraulic systems can be found on ships. Hydraulic systems on ships are used in various applications. For example, systems used for cargo systems make carrying heavy materials and performing other cargo operations easier and less time consuming.

A ship's engine room also includes hydraulic systems such as a hydraulic automatic control system. These help to regulate valve positions as well as the pneumatic air pressure in the engine room.

On top of that, hydraulic systems in a ship's stabilizers prevent the vessel from rolling and ensure a smooth performance across open waters.

Plus many industrial ships include machinery and tools like deck cranes that are run by hydraulic systems.

O-Seal Valves and Fittings and the Navy

Hydraulic systems can be found on many US Navy vessels. And with help from CPV Manufacturing and our line of O-Seal valves and fittings, these systems can ensure smooth operations and safety.

Our line of O-Seal products was developed in the 1950s when CPV Manufacturing started working with the US Navy. We wanted to make sure that every component of our high-pressure couplings met US Navy specifications. However, testing each connection would have been too strenuous and dangerous to do by hand. That's when we created a test stand using O-ring connections.

This method allows us to easily disassemble and reassemble each component to perform each test to ensure proper performance and safety. We then took those concepts and developed our line of O-Seal products.

Benefits of O-Seal Valves and Fittings in Hydraulic Systems

CPV Manufacturing's O-Seal valves and fittings are unique. Unlike other valves, our products are leak proof and designed to last. On top of that, they can withstand extreme temperatures and are rated for vacuum to 6,000 psi in liquid or gas applications, making them ideal for many types of hydraulic systems.

However, what makes our O-Seal valves truly unique is that they come with interchangeable parts. The soft goods in the cartridge can be removed and made with different types of materials for certain applications.

The versatility of our O-Seal products presents a cost-effective solution for the US Navy and many other companies across the globe. With interchangeable parts, our O-Seal valves can be used for a number of applications, which means companies no longer purchase additional valves to run their systems.

How Does a Hydraulic System Work?

You're probably already familiar with some of the basic ways a hydraulic system works. From your experience, you probably know that solids are typically impossible to squish. If you pick up a solid object like a pen or piece of wood and try to squeeze it, nothing's going to happen to the materials. They won't compress or squish. Liquid works in the same way. It is incompressible, meaning it won't squeeze when you apply pressure to it. It takes up the same amount of space as it did when pressure wasn't applied to it. Picture water in a syringe. If you cap the end of it with your finger and try to press down, neither the water nor the plunger will go anywhere.

Where hydraulics are concerned, that incompressibility is a major player in making them work. In that same syringe, if you press down on the plunger normally, you'll release the water at high speed through the narrow end, even if you didn't apply that much pressure. When you push down the plunger, you apply pressure to the water, which will try to escape however it can — in this case, at high pressure through a very narrow exit. This application shows us that we can multiply force, which we can then use to power more complex devices.

In a very simplified system, a hydraulic system is made with piping that has a weight or piston on one end to compress the liquid. As this weight depresses onto the liquid, it forces it out of a much narrower pipe at the other end. The water doesn't squish down and instead pushes itself through the pipe and out the narrow end at high speed. This system works in reverse as well. If we apply a force to the narrow end for a longer distance, it will generate a force capable of moving something much heavier on the other end.

Blaise Pascal, a French mathematician, physicist and inventor, standardized these properties in the mid-1600s. Pascal's Principle states that, in a confined space, any change in pressure applied to a fluid transmits through the fluid in every direction. In other words, if you apply pressure to one end of a container of water, the same pressure will be applied to the other side. This principle is what allows the force to be multiplied and affect a larger, heavier object.

There is a little bit of a trade-off with this system. You can typically apply more force or more speed to one end to see the opposite result on the other. For example, if you press down on the narrow end with high speed and low force, you'll apply high force but low speed to the wide end. The distance your narrow end can travel would also influence how far the wide one will move. Trading distance and force is typical in many systems, and hydraulics are no exception.

The multiplication of force is an influential factor in lifting heavy objects. If the piston in the broader side is six times the size of the smaller one, then the force applied to the fluid from the larger piston will be six times as powerful on the smaller end. For example, a 100-pound force down at the wider end creates a 600-pound force up at the narrow end. This force multiplication is what allows hydraulic systems to be relatively small. They are great for powering huge machines without taking up too much space.

Hydraulics can also be very flexible, and there are many different types of hydraulic systems. You can move the fluids through very narrow pipes and snake them around other equipment. They have a variety of sizes and shapes and can even branch off into multiple paths, allowing one piston to power several others. Car brakes are usually an example of this. The brake pedal activates two master cylinders, each of which reaches two brake pads, one for all wheels. You can find hydraulics powering a variety of components through cylinders, pumps, presses, lifts and motors.

Hydraulic systems have a few essential components to control how they work:

- Reservoir: Hydraulic systems usually use a reservoir to hold excess fluid and power the mechanism. It is important to cool the fluid, using metal walls to release the heat generated from all the friction it encounters. An unpressurized reservoir can also allow trapped air to leave the liquid, which helps efficiency. Since air compresses, it can divert the movement from the pistons and make the system work less efficiently.
- Fluid: Hydraulic fluids can vary, but they are typically petroleum, mineral- or vegetable-based oils. The fluids can have different properties based on their application. Brake fluid, for example, needs to have a high boiling point due to the high-heat mechanism it goes through. Other features include lubrication, radiation resistance and viscosity.

Let's take a look at how hydraulics typically work in heavy equipment:

How do Hydraulics Work

- Engine: This is usually gasoline-powered and allows the hydraulic system to work. In big machines, this needs to be capable of generating a lot of power.
- Pump: The hydraulic oil pump sends a flow of oil through the valve and to the hydraulic cylinder. Pump efficiency is often measured in gallons per minute and pounds per square inch (psi).
- Cylinder: The cylinder receives the high-pressure fluid from the valves and actuates the movement.
- Valve: Valves help to transport the fluid around the system by controlling things like pressure, direction and flow.

Other machines that make use of hydraulics include vehicles on construction sites. Diggers, cranes, bulldozers and excavators can all be run by robust hydraulic systems. A digger, for example, powers its massive arm with hydraulic-powered rams. The fluid is pumped into the thin pipes, lengthening the rams and, by extension, the arm. The hydraulic power behind this can be used to lift enormous loads. Aside from construction machines, hydraulics are used for everything from elevators to motors, even in airplane controls.

What is the Difference Between Open vs. Closed Hydraulic Systems?

Open and closed systems of hydraulics refer to different ways of reducing pressure to the pump. Doing this can help reduce any wear and tear.

In an open system, the pump is always working, moving oil through the pipes without building up pressure. Both the inlet to the pump and the return valve are hooked up to a hydraulic reservoir. These are also called "open center" systems, because of the open central path of the control valve when it is neutral. In this case, hydraulic fluid returns to the reservoir. The fluid coming from the pump goes to the device and then returns to the reservoir. There may also be a relief valve in the circuit to route any excess fluid to the reservoir. Filters are usually in place to keep the fluid clean.

Open systems tend to be better for low-pressure applications. They also tend to be cheaper and easier to maintain. One caution is that they can create excess heat in the system if the pressure exceeds valve settings. Another location for added heat is in the reservoir, which needs to be big enough to cool the fluid running through it. Open systems can also use multiple pumps to supply power to different systems, such as steering or control.

A closed system connects the return valve directly to the hydraulic pump inlet. It uses a single central pump to move the fluid in a continuous loop. A valve also blocks oil from the pump, instead sending it to an accumulator where it stays pressurized. Oil remains under pressure but doesn't move unless it is activated. A charge pump supplies cool, filtered oil to the low-pressure side. This step maintains pressure within the loop. A closed system is often used in mobile applications with hydrostatic transmissions and uses one pump to power multiple systems.

Open Vs Closed Hydraulic Systems

These can have smaller reservoirs because they just need to have enough fluid for the charge pump, which is relatively small. An open system can handle more high-pressure applications. The closed system offers a bit more flexibility than an open system, but that also comes with a slightly higher price tag and more complex repair. Closed systems can work with less fluid in smaller hydraulic lines, and the valves can be used to reverse the direction of the flow.

You can even convert an open system into a closed system by replacing some of the components and adding space for the oil to go after the return trip.

Types of Hydraulic Pumps

There are several different types of hydraulic pumps. These can vary significantly in the ways that they move fluid and how much they displace.

Types of Hydraulic Pumps

Almost all hydraulic pumps are positive displacement pumps, meaning they deliver a precise amount of fluid. They can be used in high-power applications of over 10,000 psi. Non-positive displacement pumps depend on pressure for the amount of fluid they move, while positive displacement pumps do not. Non-positive pumps are more common in pneumatics and low-pressure applications. They include centrifugal and axial pumps.

Positive displacement pumps can have either fixed or variable displacement. Most pumps fall under fixed displacement.

In fixed displacement, the pump provides the same amount of fluid in each pump cycle.

In variable displacement, the pump can provide different amounts of fluid based on the speed it is run at or the physical properties of the pump.

A gear pump is inexpensive and more tolerant of fluid contamination, making them suitable for rough environments. They may be less efficient, however, and wear more quickly.

External gear pumps: These make use of two tight-meshed gears within a housing. One is the driving, or powered, gear, while the other is driven, or free-flowing. The fluid is trapped in the space in between the gears and rotated through the housing. Since it cannot move backward, it is forced through the outlet pump.

Internal gear pump: The internal gear design places an inner gear, possibly with a crescent-shaped spacer, inside of an outer rotor gear. The fluid is moved via eccentricity — the deviation of the gear from circularity — between the gears. The inner gear, with fewer teeth, turns the outer gear, and the spacer goes in between them to create a seal. The fluid is drawn in, moved through the gears, sealed up and discharged.

Next up is vane pumps. These can be unbalanced or balanced and fixed or variable-displacement. They are quiet and work in pressures under 4,000 psi.

Unbalanced vane pump: This fixed displacement pump has a driven rotor and vanes that slide out in radial slots. The rotor's level of eccentricity determines the level of displacement. As it rotates, the space between the vanes increases, creating a vacuum to draw fluid in. The trapped fluid moves around the system via the rotating vanes and is pushed out as the space between them decreases.

Balanced vane pump: The balanced vane pump, also fixed displacement, moves the rotor through an elliptical cam ring. It uses two inlets and outlets on each revolution.

Variable-displacement vane pump: The displacement in this type of pump can change via the eccentricity between the rotor and casing. The outer casing ring is moveable.

Our last category of pumps is piston pumps, which are great for high-powered applications.

In-line axial piston pumps: In-line pumps align the center of the cylinder block with the center of the driveshaft. The angle of the swash/cam plate helps to determine the amount of displacement. The inlet and outlet are located in the valve plate, which connects to each cylinder alternately. As the piston moves up past the inlet port, it pulls in fluid from the reservoir. Similarly, it will push the liquid out of the outlet port as it passes it.

Bent-axis axial piston pumps: The bent-axis pumps line the center of the cylinder block at an angle with the center of the drive shaft. This design works similarly to the in-line axial pump.

Radial piston pumps: A radial piston pump uses seven or nine radial barrels, along with a reaction ring, pintle and driveshaft. The pistons are set radially around the drive shaft, and inlet and outlet ports are in the pintle, a type of hinge.

Hydraulic Systems

A hydraulic system uses a fluid under pressure to drive machinery or move mechanical components.

Description

Virtually all aircraft make use of some hydraulically powered components. In light, general aviation aircraft, this use might be limited to providing pressure to activate the wheel brakes. In larger and more complex aeroplanes, the use of hydraulically powered components is much more common. Depending upon the aircraft concerned, a single hydraulic system, or two or more hydraulic systems working together, might be used to power any or all of the following components: wheel brakes nose wheel steering landing gear retraction/extension flaps and slats thrust reversers spoilers/speed brakes flight control surfaces cargo doors/loading ramps windshield wipers propeller pitch control

A hydraulic system consists of the hydraulic fluid plus three major mechanical components. Those components are the "pressure generator" or hydraulic pump, the hydraulically powered "motor" which powers the component concerned and the system "plumbing" which contains and channels the fluid throughout the aircraft as required.

Hydraulic Fluid

Fluid is the medium via which a hydraulic system transmits its energy and, theoretically, practically any fluid could be utilized. However, given the operating pressure (3000 to 5000 psi) that most aircraft hydraulic systems generate in combination with the environmental conditions and strict safety criteria under which the system must operate, the hydraulic fluid that is used should have the following properties:

- High Flash Point. In the event of a hydraulic leak, fluid ignition should not occur at the normal operating temperatures of the surrounding components. Special hydraulic fluids with fire resistant properties have been developed for aviation use. These fluids are phosphate esters and, unlike mineral oil based hydraulic fluids, they are very difficult to ignite at room temperature. However, if the fluid is heated to temperatures in excess of 180 degrees C, it will sustain combustion. The auto-ignition temperature of most aviation hydraulic fluids is in the range of 475 degrees C.

- Adequate Viscosity. Aircraft hydraulic systems must work efficiently over a broad temperature spectrum. The fluid used must flow easily at very low temperatures but must also maintain adequate viscosity at high temperatures. The ideal hydraulic fluid will have a very low freezing point and a very high boiling point.
- Lubricant Properties. The hydraulic fluid acts as a lubricant for the pumps, actuators and motors within the system. The fluid should have anti-corrosion properties and be thermally stable.
- Thermal Capacity/Conductivity. Hydraulic fluid acts as a system coolant. The fluid must be able to readily absorb and release heat.

Hydraulic Pumps

Several types of hydraulic pumps driven by a variety of power sources can be found in aviation applications. Pumps include:

- Gear Pumps. Gear pumps use meshing gears to pump fluid. Gear pumps are fixed displacement type pumps as they move a specific amount of fluid per rotation. Gear pumps may be used on low pressure systems (under 1500 psi) but are generally not suitable for high pressure applications
- Fixed Displacement Piston Pumps. Piston pumps utilize a piston moving in a cylinder to pressurize a fluid. A fixed displacement pump moves a specific amount of fluid with each stroke.
- Variable Displacement Piston Pumps. This is the most common type of pump on large aircraft. The variable displacement design allows the pump to compensate for changes in the system demand by increasing or decreasing the fluid output. This allows near constant system pressure to be maintained.

The motive power for these pumps can be generated by a wide variety of options inclusive of:

Manual. In many light aircraft, a manual hydraulic pump provides pressure for wheel brakes or flap extension and retraction.

Engine Driven. Pumps are frequently mounted on the engine accessory gear box.

Electric. Both AC and DC motors are utilized to power hydraulic pumps with three phase AC motors being most common.

Pneumatic. Bleed air powered motors are utilized on some aircraft to drive hydraulic pumps.

Hydraulic. A Power Transfer Unit (PTU) allows the hydraulic pressure of one hydraulic system to drive a pump to pressurize a second hydraulic system without any transfer of hydraulic fluid. Depending upon the installation, a PTU can be single or bi-directional.

Ram Air Turbine. In the event of an emergency, some aircraft have a Ram Air Turbine (RAT) that can be extended into airstream to generate hydraulic pressure.

Hydraulic Motors and Cylinders

Hydraulic motors and cylinders utilize pressurized fluid to do mechanical work.

Hydraulic Motors. A hydraulic motor is a mechanical device that converts hydraulic pressure and flow into torque and angular displacement or rotation. Various types of hydraulic motors such as gear, vane and radial piston motors are available. On aircraft, hydraulic motors are most often used to drive jackscrews which can in turn be utilized to power flaps, stabilizer trim and some vertically extending landing gear applications such as found on the LOCKHEED C-130 Hercules aircraft.

Hydraulic Cylinders. A hydraulic cylinder, sometimes referred to as a linear hydraulic motor or a hydraulic actuator, is a mechanical actuator that is used to provide a reversible force in a single direction. The hydraulic cylinder consists of a cylinder barrel within which a piston connected to a piston rod ues hydraulic pressure to move back and forth. Aircraft applications include landing gear extension and retraction, cargo door operation and movement of flight control surfaces.

System "Plumbing" Components

Aviation hydraulic systems, in general, are of the "open loop" variety drawing fluid from a reservoir, pressurizing it and making it available to the various user components before returning the fluid to the reservoir. The primary components of the "plumbing" portion of the hydraulic system include the following:

- Reservoir. Hydraulic fluid reservoirs are required by most aircraft systems to provide a ready source of fluid for the hydraulic pump(s) and to contain a varying volume of fluid. This variance results from differential actuator volume (dependent upon whether the actuator is extended or retracted) and for fluid thermal contraction or expansion. The reservoir size is optimized so that only the amount of fluid needed for proper function is carried. In many installations, bleed air is used to pressurize or "bootstrap" the reservoir to help prevent hydraulic pump cavitation.

- Filters. Hydraulic fluid cleanliness is essential to proper system function. In-line filters are incorporated into the hydraulic system to remove any contaminants from the fluid.
- Shut Off Valves. Hydraulic shut off valves are usually installed at the engine firewall. In the event of an engine fire, the shutoff valve is closed to prevent possible ignition of the hydraulic fluid.
- Control Valves. Hydraulic motors and actuators have an associated control valve which is positioned in response to a manual or automated system selection such as moving the flap lever. The control valve responds to that selection by positioning to allow pressurised hydraulic fluid to flow into the motor or actuator in the appropriate direction.
- Pressure Relief Valve. In some systems, especially those utilising a fixed displacement pump, pressure relief

valves are incorporated to ensure that nominal system pressure is not exceeded. If system pressure becomes too high, the relief valve opens and fluid is returned to the reservoir.

- Hydraulic Fuses. Hydraulic fuses are in-line safety devices designed to automatically seal off a hydraulic line if pressure becomes too low.
- Accumulators. A hydraulic accumulator is a pressure storage reservoir in which hydraulic fluid is held under pressure by an external source of energy. The external source can be a spring or a compressed gas. An accumulator enables a hydraulic system to cope with extremes of demand using a less powerful pump and to respond more quickly to a temporary demand. It also acts as a system shock absorber by smoothing out pulsations. In the event of a hydraulic pump failure, the energy stored in an accumulator can provide a limited number of brake applications after landing.

Hydraulic System Redundancy

Hydraulic system redundancy is achieved by two primary means - multiple systems and multiple pressure sources within the same system.

Multiple Pressure Sources. Hydraulic systems often have more than one pump available to pressurise the system. It is quite common for a system to have one or more engine driven pumps plus one or more electric pumps. In some cases, a manual pump is also incorporated. Some systems only use the electric or manual pumps while on the ground when the engines are not operating. Others use the electric pump(s) to provide an additional pressure source during high demand situations such as gear retraction or as the primary pressure source in the event of the loss of the engine driven pump(s). When an electric pump is used as the primary pressure source, a second electric pump or a Ram Air Turbine might be incorporated into the system as a backup source of hydraulic pressure. Provision of multiple pressure sources helps to ensure that the entire hydraulic system is not lost in the event of a single component failure.

Multiple Hydraulic Systems. In many aircraft, flight control surfaces are hydraulically actuated. In these cases, multiple actuators on each surface, powered from multiple hydraulic systems, are essential to ensure that the failure a hydraulic system will not result in loss of control. In modern commercial aircraft, it is common to power the flight control surfaces from three independant hydraulic systems. The control surface architecture allows for failure of two of those systems without compromising control.

Threats

Hydraulic systems are subject to several significant threats. These include:

- System Overheat. The system exceeds its maximum allowable operating temperature and must be de-energized.
- Loss of System Pressure. Loss of system pressure can occur in two different ways; loss of fluid or failure of a hydraulic pump.
- Hydraulic Fluid Contamination. Contamination can be chemical or particulate in nature and can be caused during fluid production, by improper servicing of the hydraulic system or by a component failure.

Effects

Hydraulic system overheat, loss of pressure or fluid contamination can all result in the loss of the hydraulic system and the loss of function of those components that it powers. Fluid contamination can also result in loss of hydraulic system efficiency, fluid leaks, excessive component wear and premature component failure.

Defences

The primary defense against hydraulic fluid contamination lies in robust maintenance practices. Any fluids used to service the system must be as specified in the AOM and fluid types should not be mixed. Care should be taken to ensure that the fluid is not contaminated prior to use and that no contaminants are introduced to the system while topping up the fluid. System filters should be cleaned or replaced as per manufacturer's guidelines.

In the event of a System Overheat or Loss of Pressure, following the Quick Reference Handbook (QRH) or ECAM checklists may result in recovery of the system. If the loss of pressure was as a result a total of loss of hydraulic fluid, the system is not recoverable.

Typical Scenarios

An Airbus A330 enroute from Manchester, England to Orlando, Florida gradually lost hydraulic fluid from the Blue hydraulic system. When the Electronic Centralized Aircraft Monitor (ECAM) annunciated the system Loss of Pressure, appropriate actions were completed to secure the system. The Captain contacted Company maintenance to discuss options. Due to the fact that there was no significant loss of aircraft capability due to system redundancy, that it was less than two hours flying time to destination and that there were ample diversion airfields enroute, it was decided to continue the flight to the planned destination. The flight landed in Orlando without further incident.

Just after top of climb, a Canadian military C130 Hercules aircraft enroute from Trenton, Ontario to Winnipeg, Manitoba suffered a complete loss of the Utility hydraulic system due to a rudder actuator failure. Controlability of the aircraft was not an issue as the Booster hydraulic system also provides pressure to the flight control surfaces. The Captain elected to return to Trenton which was about 70nm to the south east of the aircraft position. An alternate (gravity) gear extension was carried out and a flapless approach and landing was made. The Auxillary hydraulic system was used to provide pressure for the brakes and full reverse propeller pitch was used to help bring the aircraft to a stop. The aircraft was shut down on the runway and then towed to the ramp.

(An announcement by the Captain of a fully-boarded Boeing 757-200 about to depart which was intended to initiate a Precautionary Rapid Disembarkation due to smoke from a hydraulic leak was confusing and a partial emergency evacuation followed. The Investigation found that Cabin Crew only knew of this via the announcement and noted subsequent replacement of the applicable procedures by an improved version, although this was still considered to lack resilience in one respect. The event was considered to have illustrated the importance of having cabin crew close to doors when passengers are on board aircraft on the ground.)

The crew of a Boeing 752 temporarily lost full control of their aircraft on a night auto-ILS approach at Keflavik when an uncommanded roll occurred during flap deployment after an earlier partial loss of normal hydraulic system pressure. The origin of the upset was found to have been a latent fatigue failure of a roll spoiler component, the effect of which had only become significant in the absence of normal hydraulic pressure and had been initially masked by autopilot authority until this was exceeded during flap deployment.)

The fracture of a hydraulic hose during an A330-200 pushback at night at Karachi was followed by dense fumes in the form of hydraulic fluid mist filling the aircraft cabin and flight deck. After some delay, during which a delay in isolating the APU air bleed exacerbated the ingress of fumes, the aircraft was towed back onto stand and an emergency evacuation completed. During the return to stand, a PBE unit malfunctioned and caught fire when one of the cabin crew attempted to use it which prevented use of the exit adjacent to it for evacuation.)

CONCLUSION

This section has been devoted to an in-depth study of how to manipulate forces by controlling pressure. With a thorough understanding of the five basic pressure valve functions, the design engineer's ability to control the interaction of forces is limited only by his imagination. Hydraulic systems, when properly designed, put tons of force under precise, finger-tip control.

In reviewing your knowledge of pressure control, you should be able to differentiate between the five basic control functions, namely: relieving, reducing, sequencing, counterbalancing, and unloading. Likewise, you should understand that, no matter how complex the function, all pressure controls operate by balancing a hydraulic force with a spring. This spring balancing is readily apparent in direct operated valve designs, but it is also the key operating principle in pilot operated versions.

In reviewing the different pilot operated functions, you will discover that no matter what the design or desired function, pilot operation always works on the principle of creating either balanced or unbalanced pressure conditions across the main control element.

More specifically, in relation to pilot operated pressure reliefs, you should know what is meant by the terms remote piloting, venting, and high vent option. You should also understand the three uses of the external pilot drain: for more stable pressure adjustments, for remote pressure control, and for load sensing.

The important points covered in our discussion of pressure reducing valves were: first, the relieving ability of a pressure reducing valve, and, second, the differences between pilot operated versions with either primary or secondary control. We also mentioned the inherent ability of the reducing valves to generate heat in the hydraulic system.

In addition to your knowledge of relief and reducing functions, you should also have a good understanding of the versatility offered by the multi-function family of direct and pilot operated valves. You should know the assembly variations and application requirements in using the valve in sequence, counterbalance, overcenter counterbalance, or unloading functions. In addition, you should realize the special requirements of accumulator circuits.

The hydraulic specialist, who understands pressure and how to control it, knows just about half of everything there is to know in his field. When he combines this knowledge with the principles of flow, the designer has the world of hydraulics at his fingertips.